Anthropology

An Understanding

Baba Pakurdheen A

ISBN: 1540356612
ISBN-13: 978-1540356611

DEDICATION

I dedicate this book to the species which calls itself wise,
submitting myself humbled before the Sublime Truth
which keeps itself obscured from the eyes which do not
seek.

CONTENTS

ACKNOWLEDGMENTS

Thank You!

Prof. Anne Vallely, University of Ottawa.
Dr. DK Bhattacharya, Delhi University.
Prof. Azmatullah, Jawaharlal Nehru University.
Prof. S Sivaramakrishnan, The Madura College.
Prof. C Sudarshan, PSG College of Technology.
Dr. Rimai Joy, Amity University.

I admit here that I hold your inspiration with or without
your consent. And promise to pass it to the indigent.

1 A SPLIT AND A SPLICE

A Split and a Splice.

'Why does Education bothers us a lot? Does our Education system yearns for a Change?'

I know that everybody must have at-least had this question once in their Life.

It has become second largest question right after our question on our existence. Almost its similar with it. From the ancient, human beings have questioned their existence. A inner voice had appeared repeatedly rising the question on our Existence.

But, most of the population had suppressed it naming crazy. Most, by the sense almost 99% of us, don't even pay attention to that inner voice. We name it crazy part of our conscience or satanic whispers or infatuation or with some other terms which suits our taste and belief. The same inner voice constantly makes question on everything. It's curious to know. Most fail to understand that the inner-voice is curious part of ourselves, curious part of our existence.

All Religions, Philosophies, Theories, Inventions and other creative ventures of this mankind had come alive only because of those who pay attention to this inner-voice.

Except some question like existence, survival etc. others are duly personal quests and changes within Individuals. These universal inner questions which most of the human population hosts are called Universal Quests.

The attempt to understand the Universal Quest of 'Who am I?' let the world to form Religions, Theories and Philosophies. In the same way Universal Quest to find solution to some existing problem invented something in-turn. To this date we have a range of Universal Quests and their Seekers. Not everyone are seekers. They suffice themselves with what they have. World is pleasant, any idea that disturb this pleasantness is tagged rebellious. But the reality we are living today was once mere ideas in those rebellious minds.

Let us come to the focus. 'Do we need a change in our Education system?'

"Why? Everything is in order. Today's Education is great. Why do we need to change it? Take birth , Have an Education, Bury the Dreams, Chase the

Scores, Enter the stream, You are Secure now, Then Die peacefully."

Yup. Is this your answer? Wait a minute.

Here we fail in describing the Security. We define our-self to be secured with some jobs that help us to earn a living. We are chasing security out of insecurity. This is life. We dwell for a living ultimately when our life itself run towards death. How come we shall live secured and settled when life itself is insecure, unsettled.

The money aimed pursuit has a good definition named Business than Education. Education cannot be built on a secure money aimed platform. It's highly drudge and insecure. Even business is highly insecure though it may have pursuit of earning Money. Every Universal Noble Pursuits has extreme Insecure platform. That's the point I like to engrave here.

Earning money with Education was a 20th century innovation. This new innovation tried to cook a recipe on Education to result in Social Security and Knowledge. But the outcome was horrible. Knowledge was sacrificed to rend security. In other words, Knowledge was molded into some shape and was shown as Ultra Modern Knowledge. People were convinced to accepted this mold as knowledge by repeated infusion. And institutes manufactured hi-end educational products for the market with young minds as the raw material. People felt the safeness it gave and started calling it the 'Real Knowledge'. The old fashioned true knowledge was now the 'Rebel Knowledge' or even 'Rebel Garbage'.

But to note here. Though naming the dung as 'Real Knowledge' be the collective behavior of the population. Still many had this Universal Question 'Is this really Knowledge?'. Someday, the falsehood gets exposed itself, that's the fact.

Every product of so called 'Real Education' had, has and will have doubts regarding its authenticity. But not everyone will set their life in pursuit of finding the truth. They don't want to rebel the mainstream with such suspicions. Yeah, its 'Real Knowledge' they surge again with the stream.

Is it really possible to seek Knowledge earning both Knowledge and Money?

Seeking knowledge will earn knowledge not money. Seeking money earns Money. Seeking Independence earns Independence. Seeking X earns X. Seeking Y earns Y. Anything that comes additional to our seek will be mere and negligible. When someone says seeking X brings Y, its false. Seeking Money out of Knowledge will be mere or null. That's false knowledge.

Unfortunately we live a split world of Science and Art. The split took place between 400 to 500 years ago when the field of Knowledge expanded vastly. The scientists and technicians live a life focusing on mechanisms and functions of the stuffs ignoring the creative appreciation while the Artists have their creative appreciation devoid of the phenomena behind. There needs to be a renaissance between Art and Science is necessary for a fulfilled environment. There needs to be perfect match between the Left and the Right brain. Those who handle to establish the medial maintenance are called the Genius. That's the reason why we admire the works of Da Vinci, Einstein and other prominent personalities.

The fuse of Art and Science will be the rounded knowledge in fact and the way future must pave its way on. And this why we felt focused on our Primary and Secondary Schooling which had a well-balanced curriculum of Art and Knowledge. And this is why we feel bored and suspicious with the single field focused Higher Secondary and University. Focusing on single field must come on Maturity. Maturity is highly Individualistic and no single group can devise it universally.

This is the problem with our system, we don't allow them to get Matured. What we need now.

A split and a splice.

A splice with Arts and Science and a split in Social Security and Knowledge. That's it. Many had backed this before. But it's not easy to do this. We require a revolution. A great revolution.

The mix of Social Security with Knowledge is the root cause. In the recorded American, Chinese, European, Indian, Islamic and other golden ages this fact is well been established. Scholars of those ages lead a life seeking Knowledge by aiding themselves social security with some trades and business. We need to train ourselves on this basis. To seek knowledge with an earning. For a golden age, we need a Split and a Splice in Knowledge. Knowledge is power and nothing else.

A Split and A Splice. Work on it.

2 NOTHING HUMAN IS ALIEN

Anthropology. One may find easy to define it by the name. *The study of Man* or *The science of Human Beings.*

It is better to stop here with the formal definition; else one may find perplexing to browse through hundreds of definitions given by hundreds of Individuals.

To define a subject one requires unfathomable understanding of that subject. As this chapter will never wait till we gain unfathomable understanding in the field of Anthropology, we are in urge to give at least a novice definition for now. Let us have a quick recap of the history of Anthropology in the urge of understanding.

It's not easy to trace out the beginning of this discipline. Since the human origin there must have been some individuals in the society involved in thinking about their existence in order to understand their own selves. But here we define anthropology as a science. What makes an approach scientific?

This is called as scientific approach. We take Phenomena, explain what causes it and predict what it may cause. By employing this approach to the approaches of Individuals in the past who dared to investigate their existence we sort out the science oriented fellows from the people being philosophical.

It starts with Herodotus the great Greek historian; he described and distinguished the people living then in Persian Empire from Greece. He established a method of differentiating other people. His approach is not totally anthropological but still we shall call it to possess some quality of being scientific.

And second we shall enter the medieval period of Islamic Golden Age. Al Biruni, the father of Indology holds all tendency of having scientific approach in his study of Human. He was well versed in Arabic, Persian, Sanskrit, Greek and Latin. His writings exhibit his detailed approach in studying other people, their culture, belief, science and other cultural and social aspects. Unlike Herodotus Biruni used direct observation and Analysis of the culture he addressed.

Third. Ibn Khaldun. 14th century philosopher of Islamic Golden Age analyzed Physical, Psychological, Environmental, Social and Economic factors that in turn had effects on rise, progress and fall of the civilizations.

These three, Herodotus, Al Biruni and Ibn Khaldun possess affinity to be called as the founding fathers of primitive Anthropology, collectively.

Soon after stepping out from their dark ages, 15th century European explorers in search of wealth had firsthand experience with other cultures and they recorded it in vivid, brief and unsystematic accounts.

We must notice that until this era, modern scientific approach has not yet begun in Europe. The printing press and Martin Luther King's protestant reformation advanced Europe in Literacy. This literacy developed science as an organized activity.

Scientific approach to the studies emerged as Europe stabilized itself with Industrial Growth. This helped philosophy to take new ventures into science with contributions from eminent personalities like Galileo Galilee, Francis Bacon, Rene Descartes and Isaac Newton.

Social Science effectively distinguished itself from Physical Science at the same time Newton was formulating his unified approach on Physical Sciences. It was John Locke, a friend of Newton who understood that the rules of science applied to the study of celestial bodies is also equally applicable to the human behavior.

"Since we cannot see everything and since we cannot even record perfectly what we do see, some knowledge will be closer to the truth than will other knowledge. Prediction of the behavior of planets might be more accurate than prediction of human

behavior, but both predictions should be based on better and better observation, measurement, and reason"

Locke writes this view on prediction of Human Behavior in his *'An Essay Concerning Human Understanding'* (1690).

Francois Marie Arouet Voltaire believed in Newton's non-religious approach to study natural phenomena. He introduced the idea of a science to uncover the laws of history. This was to be a science that could be applied on human affairs and would *enlighten* those who are governed so that they might be governed better. Then comes 18th Century; the Age of Enlightenment.

Marie Jean de Condorcet described human history in 10 stages while Jacques Rousseau who thought that in the beginning the Human Society was ideal and due to Human's invention of Agriculture and Commerce all the atrocities came into being. He proposed a Social Contract to revert human back to that Idealistic enlightened stage.

The philosophers of this enlightenment age tried to use knowledge in service of the Humankind at least by reducing its sufferings. They tried to form a religion based on science which became clear in the centuries followed.

"I believe that I shall succeed in having it recognized . . . that there are laws as well defined for the development of the human species as for the fall of a stone"

- Auguste Comte

Adolphe Que'telet, Claude-Henri de Saint Simon and Auguste Comte were early positivists who worked extensively in the social wellness of the Human Beings through scientific approach of Positivism, a new found scientific approach. The definition of Positivism is volatile to define; It changed from person to person.

Saint Simon was the person who developed this positive school of thought. He explicitly tried to found a religion based on science which botched.

Comte further worked on developing positivism as a theory to unchain human from sufferings with the new found science and like Simon he tried to make his theories into a Religion for the happiness of the Mankind. It started growing outrageous causing the later minds to detest the term 'Positivism'.

"All theories in which the ultimate standard of institutions and rules of actions was the happiness of mankind, and observation and experience the guides are entitled to the name Positive"

- John Stuart Mill

And Stuart Mill changed the word positive into phenomenal or experiential to avoid the confusion.

Que'telet named his approach *'Social Physics'*, Saint Simon called his approach *'Social Physiology'* and Comte *'Sociology'*. Thus Sociology was born alongside the Physics and other physical scientific disciplines.

Then came the Imperialist age. When Imperialistic powers of Europe rose conquering other lands of America, Africa and Asia they encountered various cultures and societies. The term imperialism changed to Colonialism as the control of other people changed to govern the displaced native poor population of the Europe into the new found lands.

With Imperialism and Colonialism came a stream of new information of cultures, customs and other data of the people residing foreign lands. To establish the superiority of theirs over these foreign societies' colonialists developed a scientific approach to justify their control over. They took information analyzed incorporating so called science inside and found Amateur Anthropology.

This amateur group of Anthropologists comprised of Zoologists, Physicians, Physicists, Philosophers and others. They studied cultures of colonized and unexplored territories, measured physical features; While ethnologists studying characteristics, behavior and differences in exist between different groups; Sociologists studying contemporary modern society. Franz Boaz who crystallized the modern anthropology was yet a doctorate holder in Physics who entered the field of Anthropology in amateur standing.

This society of amateurs paved way for professionalism. Thus Modern Anthropology was born in 19th century.

Then, came Darwin's great evolutionary theory. Middle 19th century social philosophers like Herbert Spencer proposed a social based evolutionary theory with the aid of Darwin. They declared European Civilization to be the most advanced society. And they strongly believed that every society or civilization needs to pass through a set of evolutionary phase until they acquire this advanced state. Thus, analyzing the tribal and primitive societies of the world paves way to observe the past of the Europe itself.

Lewis Henry Morgon and Edward Taylor were important figures who believed in Socio Cultural Evolution Theory. This later became Social Darwinism. By the following time, theories started pouring blending Biological evolution with Social evolution concluding a solution to suggest that socially advanced societies are of higher biological traits.

It was Franz Boas in 1920's changed the course of Anthropology making the discipline serious. Boas rejected earlier ideas of Social, Ethno Centric Evolution and insisted a lot on Research, Experience and Field Work. He defined the discipline and gave it four dimension orientation which we are aware of.

At the same time, when Boas was gaining momentum in Americas, Europe developed a new way of approach known as Structural Functionalism. These functionalists studied how institutions kept society working. Thus European functionalists helped their colonial governments to frame policies in the colonies understanding how the society functions.

Then Claude Levi-Strauss proposed structuralism theory observing common patterns across the societies. He believed that there was a basic structure which every society seems to share. His observations were laid with proofs from opposing concepts, sex, marriage, rituals etc.

Then there were Anthropologists like Julian Steward, Marwin Harris who developed economic and ecological approach in studying a society or a culture which was known as Cultural Materialism or Cultural Ecology.

And to this day Anthropology has taken diverse divisions which becomes vivid that it has made itself into a science. This is the story of Anthropology and its path in a nut shell.

So, with this we shall try to define the subject.

Anthropology is basically a science. Anthropology studies human beings and studying human beings is its ultimate objective. It studies human beings of Past, Present and Future. Anthropology stands somewhere in-between Science and Humanities. It keeps one leg in the Science and other in Humanities. It uses technique whatever helps in reaching its target. With all these ideas, let us give a novice level definition for the discipline.

"Anthropology is a discipline that studies human beings, mutually being inside and outside, by exploring their past, enquiring their present and envisioning their future by the aid of

whatever scientific or humanistic discipline or methodology that seems to benefit its study on human."

Anthropology is related with Sociology with the interest of studying human in society. It had utilized and enhanced sociological research approaches like *questionnaire surveys.* Unlike Sociology, Anthropology has not restricted itself just with studying advanced human society but had taken the idea of Sociology in studying other non-advanced human societies. It had altered the course of Sociology into its style by adding or removing certain aspects.

While History focuses to record the culture and societies of the past, Anthropology shares the same interest alongside biological aspects. Anthropology uses *archive* approach technique of History. Anthropology distinguishes itself from History in its scientific approach.

Anthropology is highly related to Psychology in many interests to understand human thinking and behavior. Thinking is the unique task which makes distinction of human from animals. So, Anthropology is much interested in knowing what human thinks as an individual and as society. Anthropology uses psychology to decode thinking of the past population on the whole while trying to form a collective idea from thinking patterns of individuals of the present. Anthropology uses *experimental* approach of Psychology.

From the launch, Anthropology is extremely allied with Philosophy and Physics. Both Anthropology and Physics share their rational, empirical, skeptical, positive and other approaches to the phenomena through Philosophy. The scientific way which physics totally grabbed itself from the philosophy stands the basic platform for anthropology to pursue its scientific approach on Humans. Anthropology uses *Hypothetical deductive model* of Physics which was developed from unifying Philosophy of *Induction* and *Deduction*.

As nothing human is alien to Anthropology, it explores divisions of every discipline wherever Human is linked. Anthropology is exclusive by its attempt to create the picture of human from all existing resources available irrespective of its category. This makes it easier to define and harder to describe.

Anthropology is less a subject matter than a bond between subject matters. It's a part in history, part literature; in part natural science, part social science; it strives to study men both from within and without; it represents both manner of looking at a man and a vision of a man- the most scientific of humanities, the most humanistic of sciences.

- Eric Wolf

Eric Wolf calls anthropology not as a subject but a bond between more than one subjects. He is trying to conclude that anthropology itself is a part in sociology, part in history, part in philosophy, part in psychology, part in social sciences, part in human sciences and a part in any other discipline which has human degree in it. And also it is an emancipated subject whose method and approach changes with advancement in time and space.

Why is it so? Why anthropology always exhibits itself to be a rebel, a misfit, trouble maker being more humanistic in sciences and more scientific in humanities?

Nothing weird. It perfectly reflects its subject, Human. Humans are of such sort. It's hard to categorize human. Change is the matter which differentiates human from other beings. Human with his rapid intelligence development makes himself a square peck to fit a round hole.

This is the evidence that Anthropology itself is a study of Human, the peculiar being who thinks whom he is. Everything will change about and so the Anthropology, I believe. The definition will remain same but description changes with advancement; that is reality.

Anthropology forms itself venturing into other disciplines continue accumulating whatever is related to Human. This is how Anthropology is defined and related.

3 JUST ANOTHER APE

Do animals have culture?

Probably everyone would say instantly 'Yes'. Google the above question, you'll find loads of articles carrying this story.

Chimpanzees found inheriting certain abilities which were not biological. Whale invents a technique which is followed by other whales. Thus Animals do have culture.

Is it?

This is the modern inclination you consider. This style is believed to be scientific. It's trend these days to believe in an article whatever it claims itself to be scientific. But the truth is most of the population doesn't know what science is. Popular article writers utilize this ignorance of their readers and they tend to pour traffic onto their sites by writing something which they claim scientific.

Let us take for instance the chimp and whale story. What's the objective behind these experiments? Were these experiments conducted with pure objective to unfold the truth of animal culture?

Definitely not. These are observations. How one shall conclude that animals have culture only with these observations whose objectives are different. Are they valid?

It seems to me that there are two types of primate studies, or two sets of writing about primate behavior; (1) those of field workers, in which the behavior is coldly measured, and (2) those of people trying to prove how much we have learned about the human behavior and evolution from the study of primates. The first kind is usually of excellent quality and cannot be seen as anything but a tremendous addition to our knowledge. The second type is often nonsense and simple bandwagoning.

- Ralph L. Holloway

This is first-class clarification for those going behind such fake scientific articles. Well, here we shall pierce the question with bursting scientific phase.

Is it so, do animals have culture?

For now, Let animals be aside. Come to culture.

What's culture?

Oxford dictionary of English utters that the word *culture* is derived from the Latin word *colere* which means *tend* or *cultivate*. In whole picture *culture* means *complex process of Cultivation with an intention of benefit*.

Agriculture is a complex process of cultivation of crops with an intention of benefit. Horticulture is complex process of cultivation of flowers with an intention of benefit. Likewise, Tissue Culture is complex process of cultivation of tissues or cells in group with an intention of benefit. Viral Culture is complex process of cultivation of viruses with an intention of benefit.

If you remove these words *an intention of benefit* from the above definition of culture, Agriculture and Horticulture will be named 'Unwanted weed growth'. Whereas, Tissue Culture and Viral Culture will be termed 'cancer' or 'disease' without the objective of benefit.

Great then, what's the point here comparing *that* culture with *this* culture. Coming. I'm trying to define culture right from its origin and evolution so that it will not look like definition but a probe.

The word 'culture', in 17th century was first time used in completely different aspect. Philosopher's started using the word 'culture' as cultivation of soul, *complex process of cultivation of good values with an intention of greater benefit for oneself and for the whole humankind* 'they defined.

And in 19th century E.B.Taylor used the word 'culture' in his book. He defines culture as 'the complex whole shared by Man as a member of society'. This is the first formal definition of 'culture' from an Anthropologist. And anthropologists developed the definition further.

Today, the definition stands saying *'culture is universally human capacity to classify and encode human experiences symbolically and to communicate symbolically the encoded experiences socially'*. If this definition is much heavier let us say precisely that *'Culture is imposition of arbitrary form upon the environment.'* This is the definition of culture by Halloway.

Culture is the ability or capacity of human beings to classify and represent experiences with symbols, and to act imaginatively and creatively.

Ok. Ease up. Let's leave everything here and shall encounter another question. *'Are humans superior to animals?'*

Most of those who said 'Animals do have culture' first, will say 'Yes' to this question with affirmation or hesitation. But no one would say 'No'. Human beings are superior beings to Animals' because this is evident. Why do we say so? We do realize that human establishes himself superior to other beings by certain attributes. Even, our discussion of other animals, their culture and behavior shows that we attempt to establish our superiority over them. This is apparent, animals are lesser or not superior to Humans.

Animals do have culture but we are superior, this is the conclusion that most people try to give. Let us investigate further ahead of wrapping up.

I'll say we are not superior.

We're not superior to a bee or a parrot or a cat in Vision.

We're not superior to a bat in hearing.

We're not superior to a Dog in differentiating smells.

We're not superior physically to an Elephant.

We're not superior to a Lion, in hunting.

We're not superior to a sloth in sleeping.

We're not superior to a bird in flying.

We're not superior to a Whale in swimming.

We're not superior to a Cheetah in running.

We're not superior to anything except in one thing that is advanced intelligent and social behavior.

This superiority or complexity of Intelligence and Social Behavior which we are highly advanced at, than other beings is called culture. *Culture is imposition of arbitrary form upon the environment.*

Though we are highly inferior to other animals in every sort, we have a capability to impose some arbitrary things on the environment. With this imposition we attempt to view, hear, smell, strengthen, hunt, fly, run, and swim in advanced manner than those animals. We impose certain strokes to encode a picture of a tree or an action and later with those strokes we decode the picture or action. That is human tendency, the cultural tendency. Still stubborn?

'We say 'Animals have culture' with only one support. Animals do inherit certain tendencies of Humans like teaching to younger generation, symbolizing unto certain extend etc. They, do inherit by non-biological process, why don't we call these tendency to be cultural?

'Man's social behavior could be compared directly with that of other species, and interpreted by the same Darwinian concepts, fruitful areas of research comparable to those developed in comparative ethology might be, for example: territoriality, optimum population maintenance, agnostic behavior, mating and consort behavior, ritualized display, play, intergroup relations, communication systems, etc. This expansion of orientation should lead to a better understanding of the non-cultural aspects of human social systems and in consequence to a sharper appreciation of the role of culture in human adaptation.'

- Tiger and R. Fox, The zoological perspective of social science.

This is a list for Social Behaviors from Tiger and Fox which still extends advancing.

The debate that 'Animals have culture' which emerges by the empirical evidence of animals to inherit by non-biological process falls completely under the above category. There is no superiority or complexity of Intelligence and Social Behavior in these observations which is necessary for a cultural aspect to arise.

Hallowell develops an idea of 'Proto-Cultural stage' to bridge this gap between Social Behavior and Culture. Further, Hockett and Ascher adds design features of *productivity, traditional transmission and duality of patterning and attributes of self-objectification, symbolic reference, self identification, self-awareness and self-appraisal* to the evolution of proto-cultural stage to cultural stage. With this they try to map the behavior of early hominids.

Likewise, Maruyama writes that *'the invention of symbolization or the capacity to structure the environment arbitrarily, is thus an initial kick which starts the process moving in the mutual-casual interplay between cultural and biological sectors of human evolution, e.g., expansion of brain, tool complexity, manual dexterity, social structure based on cohesion, communication'.*

Terrance W.Deacon writes in his book 'The Symbolic Species' that *'Biologically, we are just another ape. Mentally, we are a new phylum of organisms'*, and further he argues that language of Humans can never be compared with Animal communications since they're not analogic but vastly divergent.

He further writes, *'Though we share the same earth with millions of kinds of living creatures, we also live in a world that no other species has access to. We inhabit a world full of abstractions, impossibilities, and paradoxes. We alone brood about what didn't happen, and spend a large part of each day musing about the way things could have been if events had transpired differently. And we alone ponder what it will be like not to be. In what other species could individuals ever be troubled by the fact that they do not recall the way things were before they were born and will not know what will occur after they Die? We tell stories about our real experiences and invent stories about imagined ones, and we even make use of these stories to organize our lives. In a real sense, we live our lives in this shared virtual world. And slowly, over the millennia, we have come to realize that no other species on earth seems able to follow us into this miraculous place.'*

Again, Culture is imposition of arbitrary form upon the environment.

Let me conclude with Holloway's conclusion, *'Culture is ours alone, by the facts of arbitrariness and imposition. Logical frameworks which necessitate a priori decisions regarding the placement of an event in the past in a psychological framework will not determine the presence of "culture"; an analysis of stone tools will. A return to the essential problem, the "imposition of arbitrary form upon the environment", might serve as a stimulus for discussion that will eventually return culture once again to our own domain.'*

That's it. Human being alone possesses culture. Culture is the entity which separates human from animals. Read this chapter again if required.

4 TASTE OF NOTHINGNESS

Why do people love to do some acts while hating to do some others? How come our mind divides act of joy with sadness? Why we tend always towards joy ignoring pain? Why we divide pain and joy? Why joy takes us towards creative part and pain towards radical path? Why the path of searching joy is addictive? Is that addiction good? Then why this additive path of joy brings pain aftermath? What's the link between Joy and Pain? Are they both, both sides of same entity? Is there a path to overcome and take charge of Joy and Pain?

There are certain gawks which transcends deep into the heart. These may be a gawk of vision or sound or any other sensual strike. But they do the same. They tremble. They make us to savor the taste of nothingness. Whatever be the philosophy we go through in our life, every philosophy aims in this. 'Relishing the taste of nothingness'. When one realizes himself to be nothing, he realizes the truth. The truth is nothing but nothing itself. The things which we sense and not sense came into existence from that nothing, physics assures. This process of tasting nothing really makes us to forget whom we are and it makes us to get dissolved or lost into the nothing world.

On love, this nothingness is felt. When the love is really towards lovable thing nothingness is felt strongly. A sight, a touch, a kiss of beloved will teach this taste of nothingness. That is primitive stage in experiencing nothingness. A write-up, a poem, a song can also take us to this state of nothingness. We can induce the soul to take the nothing path thus. These are preliminary stages to taste nothingness. Tasting nothingness out of nothingness is the utmost technique. A meditative stage it is. The ability to think of 'nothing'; Making mind nil of thoughts; Taking it to the original state; to the original state of nothingness.

At this stage there is no ache, no sorrow, no complain, no misery, no sin, nothing. The original state; the nothing state. When there is nothing, ecstasy flows. Elation gushes. Celebration begins. Splendor unveils. Truth prevails.

Though above words on the process of tasting nothingness seems metaphysical; with radical approach of science we can establish it as a process of ecstatic unconsciousness due to hormone reactions. Science also empirically proves this process of tasting nothingness or getting into ecstatic unconsciousness as the mother of creation. Creativity in Human beings begins with such ecstatic unconsciousness. Science too agrees that creativity starts with nothingness.

Humans since their known history had chased this taste or stage of nothingness through various means. Through relationships, through arts, through drugs, through religions and various other means. They had tasted it; the taste of Nothingness. They had gained a lot from it; the taste of nothingness. They won't leave it; the taste of nothingness.

Since our birth we search pleasure through various stages in various names. We chase something to gain something in this life. What is the objective beyond this chase? What is the objective beyond this pursuit?

Objective is simply, Happiness. 'Pursuit of Happiness' is life's secret. Though we chase happiness, the truth is that we know nothing about happiness and that nothing itself is the happiness. We don't know what happiness is, but still we chase. We think acquiring something is real happiness. But the bitter truth is happiness is in realizing the taste of nothingness.

Let me wind this philosophic style of writing for now and let us speak scientific.

Here I like to speak about Graham Hancock's idea on Creativity and Intoxication. He believes that every culture which soared high during its time had a hand in intoxication by one or the other way. He tries to make a sketch through research in usage of intoxicants extracted from Mushrooms, Plants and other sources in various civilizations. He attempts to conclude that all mighty civilizations need creative minds and this creativity is born only through use of certain intoxicants. Intoxication gives birth to Creativity. To be creative, one needs to be intoxicated, he says.

Yeah. This intoxication is necessary to be creative. You might have noticed this simply with lovers compiling poems. The state of tasting nothingness or unconscious ecstasy can be achieved only through intoxication. And this intoxication gives birth to creativity. Let me make my point clear. How to get intoxicated?

The word intoxication makes an image in our mind as being in state of Drunkenness or intoxicated due to use of certain drugs etc. But does it stop there?

Naomi Wolf in her book '*Vagina*' describes this ecstatic unconscious state of nothingness to be related to 'Coitus'. The book starts with losing her post coital ability to taste nothingness due to minor problem in spine. The book further unfolds highlighting Vagina to be a highly creative organ rather than being sexual. Precisely, she attempts to portray that one can enter the ecstatic unconscious stage of tasting nothingness through coitus.

From, Hancock and Wolf, one can taste nothingness through Drugs and Sex, is this conclusion?

Indirectly, west tries establishing that Drug and Sex is behind their success. Since Atheistic values are today's scientific trend, metaphysical considerations are disregarded. You must have noticed first half of this article to be more metaphysical. Scientific minds must have just skipped the chapter declaring it to be less scientific. But do you know a fact that this metaphysical poetic expression was the language of scientific accounting for centuries. Can we ignore some ideas which are not written in so called scientific style?

Let us take account of some creative personalities of the West. Be Darwin, Einstein, Mozart, Newton and Picasso. Did they achieve their creative skills through Drugs and Sex? But they were creative, how? Weren't they tasted the ecstatic unconsciousness of Nothingness?

Yeah, each and every creative brain tastes the nothingness. They did. They even tried to explain their experience of tasting nothingness in metaphysical style. Note it down, even scientists like Newton and Einstein who spoke in language of science tried to explain this feeling of tasting nothingness through Metaphysical language. Why?

"In fact, science does not reject metaphysical knowledge— though individual scientists may do so—only the use of metaphysics to explain natural phenomena. The great insights about the nature of existence, expressed throughout the ages by poets, theologians, philosophers, historians, and other humanists may one day be understood as biophysical phenomena, but so far, they remain tantalizingly metaphysical."

- Russel Bernard, Research Methods in Anthropology.

As a student of Anthropology one can't ignore anything on pretext of the way it has been said since our task is to be more scientific in artistic approaches and more artistic in scientific approaches. When science bounds, Metaphysics takes breath.

Here ceases west. West tries to restrict here by concluding that creativity begins through Drugs and Sex. There is no further research in this subject. I feel The East on this subject has developed further research though they may seem metaphysical. This research of the East to find a way to be in the state of ecstatic unconsciousness can be found to be included in aspects of all Religions and Philosophies. All formalized religions such as Buddhism, Christianity, Hinduism, Islam, and Jainism are from the East. East made continuous efforts to find a way to taste nothingness.

East divides clearly this state of intoxication or the state of being in ecstatic unconsciousness tasting nothingness into two dichotomous categories. Minor and Major approach. In other words Temporary and Permanent approaches.

East declares Drug and Sex approach of attaining ecstasy as temporary or minor approach and Inward-Outward quest to be Permanent or Major approach. Knowingly or unknowingly every creative mind has taken this inward-outward quest technique to make itself creative.

Aftermath Drug or Sex, mind and body experiences certain degree of after effects. And the ecstasy just vanishes in few hours. But Inward-Outward quest leaves the mind in the state of ecstasy forever. This is the reality.

Every creative personality whether he is from west or east has taken this permanent approach of Inward-Outward quest to taste nothingness.

Though Eastern Philosophies may seem unfit for science to probe, in truth it is not. There is certain degree of truth which one could sense. For now they may have only metaphysical or philosophical answers. But recent path of science shows how our metaphysical assumptions are nearly truth.

Here are the questions again. We know well that we've not been answered straight. Let's Practice ourselves by surging our brain to bring answer for the questions.

Why do people love to do some acts while hating some? How come our mind divides act of joy with sadness? Why we tend always towards joy ignoring pain? Why we divide pain and joy? Why joy takes us towards creative part and pain towards radical path? Why the path of searching joy is addictive? Is that addiction good? Then why this additive path of joy brings pain aftermath? What's the link between Joy and Pain? Are they both, both sides of same entity? Is there a path to overcome and take charge of Joy and Pain?

Proper approaches or developments to taste nothingness will unchain humans from all addictions and will heal. We shall know the truth behind Joy and Pain.

Joy, Pain and Taste of Nothingness.

5 THE NOBLEST TASK

Feeling bored? Get prepared for a grand task. A task which you will love. A task which humans love.

What's the task that humans love a lot? Have you ever had this question?

What's the task that humans love a lot?

Take few minutes to think.

While we were children we used to be curious. Curious to know. Curious to know mechanisms behind our surrounding. We were curious and daring enough to investigate the mystery we saw all around us. We were curious. Our curiosity promoted investigation.

Human being in full circle is a curios being. Curiosity to see tomorrow made him human. He was curious and daring enough to explore the world around him. Curiosity itself is not a task but a quality. What task will curiosity lead to?

Investigation. Curiosity leads to investigation.

He was curious. He was daring enough to act on his curiosity. He utilized his maximum physical and mental ability to investigate what he was curious about. When his ability restrained him, he invented. He invented external aids to enhance his investigation. He thus became Human.

Human investigation forever instigates extrinsic. We spot error in others prior to us. We set up our childhood thoroughly investigating the fresh planet around and its peculiar phenomena. We attempt to discover answers by hearing the explanations for phenomena from elders those who have experienced. Then we raise questions. When our radius extends further we hear new explanations, which cause constant conflict and settlement inside us. We analyze carefully the new stream of ideas into our lives. We employ our considerations upon it. We make assumptions if we don't know or like or understand the explanation provided. The cycle continues. Thus we build our self with investigation. Thus we are in continuous form of learning throughout our life by investigating.

Take some time and understand the passage above. This is what we love doing though may feel a bit tough to read.

Although investigation begins extrinsic, it probes both in and out. It builds 'self-in' with our own ethics, values, assumptions, and attitudes etc. keenly investigating out. Also in reverse by inward investigation like imagining and thinking it builds 'world-out'.

This is the noblest task. Scientifically *'Investigation'*, Philosophically *'Search'*.

Investigation is the task following every pace of human life. Science and Philosophies are just outcome of human investigation.

Human started investigating his surroundings and himself from the very beginning day he was known human. He correlated all his investigations (hearings, understandings, observations, assumptions, conclusions etc.) and gave theories. Philosophy was thus born. Those who gave theories after investigating were called philosophers.

Philosophical theories have no bounds. They work on approval and disapproval of group of philosophers. When one's investigations are in phase with other investigators they are approved, otherwise disapproved. Approved doesn't mean that theory is True and disproved doesn't mean that theory is untrue. Approved and Disapproved on the pretext of others, that's what it mean, not concerning truth. The theories rejected may be understood by forthcoming generations and may be approved later. Time do played a crucial role in theories.

Then, Philosophy got advanced with time. Philosophy investigated investigation's nature and devised certain criteria for investigations and turned Science. Simply, science is a well formalized investigation. Science became science by investigating investigations. The word *'investigating investigations'* is a solid word which needs separate article to explain. Please take enough time here and investigate yourself the sentence 'Investigating Investigations'. Every discipline emerged from philosophy including science. And philosophy ultimately from the task of Investigating.

In our times we are witnessing vast division between Arts and Science, though both being emerged from Philosophy. We have discussed already the need for fusion between Arts and Science in chapter *'A Split and A Splice'*.

In the beginning I told we are going to do a task, what's that?

We are going to Investigate. What are we going to investigate?

We are going to investigate the beings which investigate. We are going to investigate ourselves.

'Who are we? '. This is the objective.

It is hypothetical question for which numerous thinkers of the past had tried to give an answer.

With advancement in Philosophy we are trying to investigate this question. Not scarcely in strict scientific manner but also in Humanistic manner.

Our objective is one, 'to study human beings in holistic manner'. We are going to investigate human beings of Past and Present irrespective of region. We are going to investigate humans in all frames of time and space.

Our objective is singular to study ourselves. So, no matter which method we choose to investigate. We would choose any method that suits our investigation irrespective of its scientific or humanistic origin. We will feel no disgrace to enter any field which deals with us. Either it is History or Psychology or Medicine or Zoology or Economics or Politics or Engineering or Physics or any other known or unknown discipline. We feel no humiliation to take our share from them.

Our objective is this, 'to study ourselves'. If there's an approach to reach our goal in other disciplines we will follow. If not we will alternate it or invent a new approach.

If Humanistic approach seems more humanistic let us soak it in scientific bath. If scientific approach seems more scientific let us soak it in humanistic bath. We will make a more scientific humanistic discipline and more humanistic scientific discipline.

But keep in mind that this investigation is harder than we think. The path for this investigation is not at all trodden. It merely has an uneven trail here and there. We need to climb on rocks, hang on ropes, clear routes to reach the destination. We need to leave a well-connected trail for the forthcoming generation to trod roads. We need to find a way to Balance Humanities with Science, Right with the Left, Creativity with Radicalism, East with West, Good with Bad. It's possible. Believe.

Are you ready for this investigation, to investigate the being which investigates?

Anthropology begins.

6 THE DUAL DIMENSION

Investigation of Human Beings.

Title seems great. How to investigate? We have comprehended an outline of the investigation and its necessity in last chapter. Now, how to investigate?

I'll come straight to the point. Human beings have a unique quality which no animal possesses. It's completely unique. The dual nature. The dual nature of human beings is that unique tendency. This is dual nature of human being to act in dimension of Physical and Cultural frames as two separate being. Human Being is combination of Physical and Cultural dimension of being.

Can we relate dual nature of human being to hardware and software in computer language? Since, Physical Dimension of human to hardware as they could be observed vividly and software to Cultural

Dimension as they can only be felt not seen. Can we relate?

No we can't. The Physical and Cultural being are the apt terms to describe human dimensions. The terms 'Hardware' and 'Software' may be synonymous with 'Physical' and 'Mental' dimensions. Not with 'Physical' and 'Cultural' dimension. Animals do have the duality of Physical and Mental dimensions. In physical dimension human is merely another animal but mentally he is not. Mentally he forms a new array of species. He has unique quality which clearly differentiates him from other animals. We call that higher mental dimension of human being as 'Culture'.

The term 'Cultural Dimension' is specific only to humans as animals don't possess culture. Read *Just another Ape'* to settle for a conclusion.

So, human exists in two dimensions, Physical and Cultural. And both of these entities highly depend upon each other and are not independent. Both have their influence over each other to certain extent. Humans change their Cultural Dimension in accordance with their Physical Dimension and vice versa. We will discuss later how they're interdependent. For now let us surf Physical and Cultural Dimension.

Physical Dimension of Human Being includes his body and all other physical products which are fashioned by him. And Cultural dimension includes all his exclusive mental dimension and its products.

Ok. Let us turn to investigation again.

How to investigate Man?

Investigate his Physical and Cultural Dimension.

How to investigate his Physical and Cultural Dimension?

Well. Consider that we are given a task to scientifically investigate you. How shall we investigate? First we shall investigate your Physical Dimension and Cultural Dimension.

We shall investigate your physical dimension by measuring your anatomy or by scientifically observing your external and internal physical features or penetrating more advanced with your DNA etc.

So we can scientifically investigate your Physical Dimension and arrive at a decision that you're measuring 6 feet in height, 67 kilograms in weight, you are dark skinned, with a prominent forehead and sleek hair. Internally your bone marrow is 70 centimetres long. Your blood group being O positive. You have diabetes. By exploring your genes we may determine that diabetes or arthritis was common with your forefathers too. Thus we can make a conclusion with our investigation on your Physical Dimension.

Similarly, we can investigate your Cultural Dimension by devoting some time to be with you or by psychological observation, questionnaires and investigations etc.

With our investigation of your Cultural Dimension we can draw up supposition that you are an Atheist, with optimistic values. Your interest ranges extremely from flirting to reading. You're philanthropist. Accordingly we can settle with your values, behavior, tastes, interest, beliefs, etc.

So, with our investigations we can conclude your whole being as Human. That simple it is.

Now let us take another case. Consider that you're interested in reviving your grandfather's life. You're interested in investigating your Grandfather who is not alive now.

How will you investigate this case? The person who is not alive. Let me spin the story further.

You are enquiring about Grandfather to your Dad, Mom and other persons whomever you feel related to him. Then with assistance of them you are gathering information about your Grandfather's Physical (his height, weight, appearance, diseases etc.) and Cultural Dimension (profession, interests, beliefs, values, behavior etc.). Now you have grouped first-hand information about your Grandfather which may or may not be true. To eliminate falsehood from truth you are switching your investigation, Scientific. You are now searching for scientific evidences and proofs to aid the first-hand information.

You are observing skeleton of your grandfather digging up his grave. You've now proofs and evidences to support his anatomical descriptions.

You're analyzing DNA remains from his bone and you're collecting scientific information as much as possible. Then, you are reconstructing your grandfather's physical dimension accordingly.

And what about the Cultural Dimension?

What your grandfather was interested in? What was his values and beliefs? How he behaved?

The persons whom you enquired for your first-hand information might provide answers according to their views on him. As your grandfather must have exhibited different sides of his persona to different people, so it will surely vary. If we appeal for a conclusion simply with first-hand data, it will never be scientific. And we can't use the same scientific investigation method used for proving physical dimension. And ultimately it will be sensible idea to leave your grandfather's skeleton to rest in peace rather than questioning it about its culture.

What shall we do then?

Remember, the interrelation between Physical and Cultural Dimension. So, third we can make an investigation on this point. We need to find a loop hole to construct Cultural Dimension with available Physical products.

Now, you're getting some physical products from your dad which your grandfather had used. Collection has some diaries and material products. By decoding your grandfather's diary you'll get to know his specific

Cultural Dimension with authenticity. And by analyzing the left material products you can assume their function and can relate them to your Grandfather's cultural dimension.

So, assumption also plays a great role in scientific investigation. Cultural dimension can never be more radical or empirical as physical dimension, but it works also on established assumptions, theories and other advanced scientific research methods.

After all this, we will be able to reconstruct your grandfather's life to certain promising degree.

Now we shall take a new task. Investigate life of the first human?

Before going further in, first we must know the importance of knowing 'who we are?' and why it's necessary to know ourselves? What will be the hope for investigating ourselves? Is it necessary to investigate past human lives? Why?

Investigation will endure.

7 THE SEAL OF ANTHROPOLOGY

Who we are? Why it's so important to know who we are? What's the objective beyond in knowing ourselves? Is there any benefit or hope for this *'we'* investigation?

Human mind always asks this question of benefit. What will we have in end if we do this or that? Many of us may have this question. Why do we need to investigate ourselves? Investigation of self is possible only by sages in fables of the past? What is the hope for this investigation? What will be the benefit?

We'll have a journey to find answer for these questions, since I believe holding doubt is a grave hindrance for progress of any task. When doubts persists, either we must clear ourselves out of it or out of the task itself. There is no way to pursue the task clinging to doubt, it will certainly make fuss.

Ok, let's begin.

In the beginning, we saw investigation is of two types; intrinsic and extrinsic. That is 'in' and 'out'. Again I replicate clearly that our task is to investigate 'in'. That is to investigate the investigator himself. Investigating 'out' implies investigating the surrounding.

If we are really able to find the gain behind investigating 'out', we will successfully be able to relate it with investigating 'in'. I've told the answer.

Why we are investigating 'in' and not 'out'?

Why we are investigating ourselves not our surroundings?

It's clear that the field of science which investigates our surrounding had grown in great range. On the other hand the hardest and noblest task of investigation ourselves has not yet grown to that range. And most of public are unaware of its importance.

Though we are beings of 'in' and 'out', we focus much on 'out' which is quite apparent. This apparent quality of the 'out' make us easy to believe its reality; whereas subtle, hectic and hidden 'in' makes the task harder to investigate and believe.

So, we pay greater attention to investigate apparent 'out' than hectic 'in'. This is human tendency. We are afraid of our hectic 'in'. But not all, it is apparent that those who are reading further are of that rebels who always like to risk in unknotting hectic mysteries.

Name three achievements of Science in your choice. Keep this question to yourself I'll exhume later.

Ok. A question again? What's purpose of knowing *'out'* or our surrounding?

Concept will be quite clear and straight forward still answers may vary. To survive first and next to control. Without knowing or investigating *'out'* we cannot survive. Since survival is the key to exist or to be in reality it is mandatory to know the phenomena behind it through continuous investigation.

When survived, control surprises. After knowing the phenomena behind the *'out'* we rush to take control. We assume ourselves the title to govern the phenomena with our wishes, which we are fully aware of. Even though all animals may have this tendency, human leads.

Name three achievements of Science in your choice.

Whatever achievement you may list. Electricity or Computer or any other achievement. You are not representing the direct achievements of Science but achievement of Technology or Applied Science. If really you need to list achievement of science your list will have Newton's calculus, Einstein's theory of relativity, Rutherford's radio-active theory and so on.

The collective conduct with public is that they measure science with aid of technology or its application. Most don't know what relativity is, though they enjoy the technology developed on that basis. Always, when science discovers something new it would seem hope-less or benefit-less for the public until it reflects in a technological benefit or application. Human will demand to feed poor in place of wasting money in mars probe which might seem rationally truthful.

This is human, they demand profit. The point is that even sometimes the investigation on apparent *'out'* seem to have *'no benefit'* in human term. Human Beings on the whole don't see the vision. The visionaries are always mere. Still, it is always binding to these mere number of humans to explain their vision to their Human spare. It is a duty too.

The pen with which I'm writing, the paper which you are reading and the dispersed ink in it, the chair on which you are reclining, the dress you're wearing, the computer's screen which you're watching, the internet which you're browsing and all other visible and invisible stuffs surrounding you (i.e *'out'*) were once metaphysical. Our investigation of the *'out'* gave us the power to survive and control it. Now, we are in charge of control. Though human may be inferior in physical terms to a whale in swimming but investigation had made us superior to them by inventing submarines.

So, investigation provides us knowledge over the thing which we are investigating, which in turn can be controlled and utilized for our benefit.

Here take a break to think about the question.

'What is the benefit of investigating ourselves?'

Are we trying to survive ourselves? Are we trying to control ourselves?

Yes. Indeed we are. We are trying to survive ourselves. We are trying to control ourselves.

Human is both a Physical and Cultural being. In a sentence, *we are aiming to survive and control the phenomena of our physical and cultural existence by investigating ourselves.*

Well. The objective with investigating *'out'* and *'in'* is clear. To survive and control. By investigating 'out' first we had learned to survive, then with excess learning we had learned to control.

Likewise, our attempt to investigate our physical *'in'* being has contributed a lot in Biological Science. We have successfully found the physiological phenomena, decoded DNA, still researching the brain. We are trying to control our physical being through achievements.

People understand importance or the profit of investigating Physical Science (*out*), Biological Science (*in*). They can list some achievements of those sciences by their application. Though latest research like quantum mechanics, cold fusion, exploring cosmos, researching brain, mutant foods, may have less acknowledgement from the public today. But when they exhibit their application they turn the scene.

The problem which people don't understand comes with our 'Cultural Being'. People can't understand what's the profit in Social Sciences? Especially Anthropology.

Though purpose may be clear, what's the use?

'Religion is Evil'. Its today's trend to say. Is it? Is religion is evil?

People throw empirical evidence of Europe's dark ages and Modern Islamic radicalism as proofs to base this theory. On the surface it may seem sure. One may conclude that embracing Science is best alternative to Religions. Is it true? Is really religion evil? Is really science noble?

We can't conclude anything like that on European standard. Europe might had suffered from religion; but not all the parts. India, China, Middle East had culturally flourished right after religious upsurge of Vedanta, Buddhism and Islam respectively. It's evident that Islamic Golden Age was promoting science and philosophy while Europe was in dark. Today's instability of Middle East is not due to Islamic radicalization but purely due to western intervention. Robert Kaplan explains the role of boundary in politics and nation building in his book

'Revenge of Geography' which explains other elements for a country's instability. Further, Russell Bernard writes in his 'Research Methods in Anthropology'

'Science did not cause Nazi or Soviet tyranny any more than religion caused the tyranny of the Crusades or the burning of witches in 17th-century Salem, Massachusetts. Tyrants of every generation have used any means, including any convenient epistemology or cosmology, to justify and further their despicable behavior. Whether tyrants seek to justify their power by claiming that they speak to the gods or to scientists, the awful result is the same. But the explanation for tyranny is surely neither religion nor science.'

Though on the surface, the question 'Is really religion is evil? Is really science is noble?' may seems to weigh more on the side of science, but ultimately it's not. This is anthropological approach to a problem. The holistic approach.

Now say me. In the beginning we thought religion to be evil but investigation has made it easy for us to remove the false notion behind the holding. That is the purpose of investigating our cultural being. It can avoid misunderstandings. It can avoid wars restoring peace. It can find a solution for poverty, hunger, racism, castes, pollution, relationships, violence etc. It can change the world, by changing minds.

This is the seal of Anthropology.

8 PECULIAR YET PROMISING

Why we need to study Anthropology?

Before stumbling upon our question, let us first recognize the necessity for study. What's the objective of study, any study?

We study to become aware of. We become aware of ourselves and our surroundings since the time we were born, through rigorous study. The study which starts naïve attains a mature, formulated state with age. Through Trial and Error we apply our study and learn from it continuously. This is how we could understand study in humblest style.

We don't study to score high marks. We don't study to get a good job. We don't study to earn money. The study of gathering money is well known as 'Business'. We simply study to become aware of. Place and Possession are mere attachments for this process of becoming aware of. That's it. We study to become aware of.

Why to study Anthropology?

Anthropology. The science of human beings in all frames of time and space. We hold adequate primer to the subject. Nevertheless, most of us don't know its importance.

If we have been aware of the need to study Anthropology we would have certainly not asked this question before. Will anyone dare to ask why to study Physics? Why to study a Language? Why to study Math?

Will anyone attempt to ask such question? Never. Why?

They know very well about their objective. They clearly know the objective behind Physics, Language or Mathematics. We ask why to study Anthropology since we are ignorant of its objective. Or we consider its importance to be trifle.

Here, I give a statement. *The need to study Anthropology holds more importance than any other discipline and it essentials to be studied right from the foundation of formulated education slope'*

Anthropology is an attempt of Human to understand himself. It's just like all philosophies but differs by its scientific approach. There it lies with a solution which we humans face. Most problems in our society arise due to misunderstanding. Here, the point is that Anthropology is subject which needs to be familiarized in primary level. Like Languages, Mathematics and other sciences anthropology is the discipline which fulfils potentials essential to be announced at kindergarten level. Children must be educated to understand themselves and those around, before entering something further.

A child grows assimilating culture and conduct by inquiring surrounding. That is from the Culture a child starts its learning process with or without consciousness.

It is culture by which a child is given identity. It is culture by which a child is bound within a society. It is culture by which a child is segmented inside its society. It is culture by which a child develops its behavior and conduct. Culture is vital.

Behavior or Mental set up of a child is chastely cultural and merely individualistic. A strong evidence to this point can be provided from the current course change of Psychology into Socio-Cultural approach from its earlier Psychodynamic, Cognitive schools. Psychology which studied behavior of Human Being which was Individualistic in the past had taken an Anthropologic approach of studying an Individual as a part or a reflection of a society or a culture. In present day situation the borders between social sciences have successfully been collapsed into the vastness of Anthropology. And Anthropology maintains its unique tendency of being borderless.

So, culture holds a great deal in existence of Human Beings. In other words Human Beings are the active medium for the culture to enact. By studying culture we study the active part of Human Being. By learning culture we shall become aware of the conscious and unconscious beings of Human.

Simply, before we teach science to understand numbers and alphabets, we need to teach them to understand who they are. Why others are different. While submerging the children into a culture we must make them to understand what it is and get accustomed to other's cultures.

All great personalities had achieved their status only by self-questioning their existence. It is more important to understand oneself than the surrounding. The study of Anthropology will do magic. It will solve every problem that arises in society of Mankind due to misperception between various institutes of culture. It breaks the border. Unites Human under their diverse term. Peculiar yet promising.

9 UNTIL THERE WILL BE MANKIND

Unlike Sociologists or Political Scientists or Psychologists or any other Social Scientists, Anthropologists are called 'The Mild Breed'.

All social science disciplines study individuals and societies in different norms with different approaches, and different research methodologies. But Anthropology as a discipline has been accused for many years for not being active as others.

Let us take example of Sociology, the discipline known for giving fierce social warriors out who constantly struggle against the problems and atrocities in society and demand for reforms and restructure. Many renowned social activists have sociological approach to social problems. And they are branded standing vanguard against any social issue.

And it's same with Political Science, Economics, and Psychology etc. They are known. They are known very sound in public domain.

Try to conduct a questionnaire in public, sampling people representing various sets of society with a simple and single question *'What is Anthropology?'*. It is assured that you will have amusement while evaluating out the responses.

Many will struggle even to pronounce the word *'Anthropology'*. Some may suggest it to be the study of stars, cockroaches or any word that raids their memory at that instance. Close social discipline colleagues will suggest Anthropology to be study of stone tools, tribals, skulls and so on. And naive Anthropologists will confidently say Anthropology is scientific study of human beings in whole which includes analysing their physical and cultural being in present and past. Don't ask them further since that's all they have been made to get.

Okay. Let it be. Do same research with questions like this 'What is Archaeology', 'What is Psychology?', 'What is Sociology?', 'What is Economics?'

Its assured that you will receive prompt replies perfectly explaining the fields and their ideas. The answers will be clear cut in comparison with prior responses to Anthropology. Anthropology is not so popular. It's true. And there remains a great confusion in understanding it.

And further from time to time in US, politicians had deliberately demanded to stop National Science Foundation's (NSF) Funding Anthropology, questioning *'Where the Science is in Anthropology?',* And recently federal government had considered this proposal so seriously leading president of American Anthropological Association (AAA) to draft mail to its members requesting coordination.

Anthropology has been accused *'objectiveless'* in nature surviving by scavenging or parasiting on other social disciplines. It's also blamed that Anthropology deludes by stating objectiveless or unattainable objective that it has holistic approach on scientifically studying human beings.

Established institutions and organizations had tried their best to open up an applied rank for Anthropology ranging from Biological to Social extents. But still a biological Anthropologist working on genetics turn out to be geneticist. An Anthropologist working on Languages becomes Linguist. An Anthropologist employed on society changes to be Sociologist. An Anthropologist working on primates converts to be primatologist. An Anthropologist detailing a tribe becomes ethnographer. An anthropologist working on past fossils and artefacts befits into Archaeologist. An Anthropologist working on something becomes that thing and not labelled *'Anything Anthropology'.*

Is Anthropology not a discipline? Does it really bugs on other disciplines? Is there any prerequisite for Anthropology when every other fields exist independently even without Anthropology? Does it certainly has an application? Where it's serious registering its existence? Does it really exists?

I'll say in a sentence, the answer for these questions before further explaining the stuff.

'Anthropology has ever been and ever will be an Approach and never a Discipline'

In both the sense Anthropology is a bit unruly in nature which can never be tamed into discipline just like its subject Human. It cannot be restricted with mock borders which divide other disciplines. The attempt of many institutes and organizations to jar it exclusively as a structured discipline even after trying to *'apply'* it commercially, has ended up in vain.

Since creation of its base by amateur travellers, merchants and mercenaries it has been an attempt to bridge lucid artistic accounts with rigid scientific recording on Human Beings. It walks peacefully like a cat on the wall with Science on one side and Arts on the other. Though this cat may look unsteady, by nature it is fully steady to handle its unsteadiness. It has both tendency to be as wild as animal and as gentle as mystic. The prudent lies with those who like to walk on its path.

Anthropology attempts to study the utmost complex, utmost chaotic, utmost mysterious, utmost enigmatic, utmost daring and utmost fluctuating existence in the known dimensions, The Homo Sapiens. The objective of Anthropology looks deceiving and unhopeful because of its most complex, most chaotic, most enigmatic, most daring, and most fluctuating objective. Which can never be wholly understandable. That why Anthropology is called 'Holistic' not 'Wholistic'. When you declare you had understood, it will change its form and nature. Anthropology in reality can never learn Human in complete, but can approach to learn him in complete.

Anthropology is not alone. It still has a great and world renowned companion with its nature in Natural Sciences. 'The New Physics' which has successfully emerged in recent decades has the same peculiar qualities of the Anthropology with its weird Standard Model, Multi Verse, Super Symmetry and all once called Mystic views. The new era of science has begun. The strict disciplinary boundaries will disappear in near future for the intelligent leaving the old trail of boundaries for applied professionals to be trained. We need to advance further to appreciate the dimension of knowledge and its nature. And permanently the matured are known for their Gentleness and are called 'Mild Breed'. They know the eventual truth. That nothing can be deduced with firmness. Nature is Inquisitive.

Anthropology though named some hundred years aback, has been in approach from the very beginning of humankind by the name of Legends, by the name of Prophecies, by the name of Poetries, by the name of Religions, by the name of Art, by the name of Science. So, attempt to remove Anthropology noting it's peculiarity will merely remove only its label yet will

be existing in some other label retaining its form and
nature, until there will be mankind.

10 BEING HUMAN

Panic besides Peace,

Solitude besides Intimacy,

Anonymity besides Eminence,

Vigor besides Infirm,

Sense.

Insecure is life,

Secure in delusion.

The illustrator of the obvious,

Is illustrated obscure.

Ache for a union.

A union,

To conclude solitude,

With spiritual and animal quest.

Will rise realizing eternal truth?

Or will grow soaked assuming falsehood?

Trapped amid Spirit and Sprite.

Being human is.

ABOUT

Baba Pakurdheen A, aka **abdheen**, is primarily a student of Anthropology and Physics who believes in integrated approach to science and advocates need for Anthropology to be introduced at primary level. This book compiles his essays on Anthropology as he understands the discipline.

Please visit author's personal site further.
http://abdheen.com/